FOUNDATION LEVEL 1

Collins

LITTLE LEARNERS

MATHS

SERIES EDITOR: RANI T.

Published by Collins Learning
A division of HarperCollins*Publishers*
India Private Limited

HarperCollins Publishers
A-75, Sector 57, Noida,
Uttar Pradesh 201301, India

1 London Bridge Street, London, SE1 9GF,
United Kingdom

2 Bloor Street East, Toronto, Ontario M4W
1A8, Canada

Lvl 13, 201 Elizabeth Street (PO Box A565,
NSW, 1235), Sydney NSW 2000, Australia

195 Broadway, New York, NY 10007, USA

Browse the complete Collins catalogue at
www.collins.in
For any queries/feedback, please write to us
at **SCHOOLSUPPORT@harpercollins.co.in** or
call us on **0120-4044800**.

First published in 2020

Reprint: 10 9 8 7 6 5 4 3 2 1

ISBN: 9789353570330

Artwork and layout by
Qualibre Creatives Pvt. Ltd.

Printed and bound by MicroPrints (India), New Delhi

Acknowledgements

Cover: HarperCollins*Publishers*India

Title Page: HarperCollins*Publishers*India

Prelim Pgaes: radenmas, denkcreative, nonohana, graphic-bee

Inside: Page 8: terdpong pangwong, Maria Voronchikhin, vectortatu; **9:** graphic-bee, Sudowoodo, vectorsmarket, luplupme, vectortatu, 7romawka7, AldanNa, fireflamenco; **10:** graphic-bee, Naddiya; **11:** IconicBestiary, saenal78, FaberrInk; **12:** dodo4466, Yayasya, StudioBarcelona, hermandesign2015; **13:** studiogstock, Eloku, luplupme, AlexeyevaJulia, terdpong pangwong, nataka, KanKhem; **14:** KanKhem; **15:** graphic-bee, colematt, jehsomwang; **16:** vectorsmarket; **17:** artmoryachok, Olga_Mallari, incomible, Oleksandr Khoma, Alxyzt, Katsapura, Redlio Designs, neyro2008, Vectorpower, Electric_Crayon; **19:** Godruma, Alexey Bezrodny, ksana-gribakina, Sky_melody; **20:** terdpong pangwong; **21:** Tomacco, D_A_S_H_U, DmitryMo, MilaArt, XonkArts, kbeis; **22:** Andrea_Hill, AlfaSmart, Bigmouse108, Vectorpower, sabelskaya; **23:** graphic-bee, sabelskaya; **24:** Oksancia; **25:** PinkPueblo, MicrovOne, Rhoeo, MariaTkach, Naddiya, incomible, Vect0r0vich; **26:** luplupme, colematt, graphic-bee, nyamol, S-S-S; **27:** eliflamra; **28:** Mr.Timmi, krishna pethani, MaksimYremenko, eduardrobert; **29:** drmakkoy, djvstock, belonogov, Mykyta Dolmatov, BOLDG; **30:** Marina Vol, daz2d, Jaswinder Singh, Bienchen-s, MaksimYremenko; **32:** colematt, yelet; **33:** Yayasya; **34:** colematt, yelet, jagadeesha Gokhale, Iuliia Kanivets; **35:** dodo4466; **36:** colematt, yelet, FloWBo, graphic-bee; **38:** yelet; **40:** colematt, yelet, FloWBo; **41:** colematt; **42:** kbeis, Mochipet, Tigatelu, TopVectors; **43:** Chuhail, Mary_stocker, neyro2008, Tetiana Lazunova, Naddiya; **45:** Hein Nouwens, Nikiteev_ Konstantin, katrink03, katrink03, katrink03; **46:** sabelskaya; **47:** blueringmedia, Olga Andreevna; **48:** colematt, yelet; **49:** colematt; **50:** FloWBo, yelet; **51:** musri; **52:** FloWBo, yelet, graphic-bee, colematt, dumayne; **53:** AKARAKINGDOMS, totallyjamie; **54:** yelet, graphic-bee, colematt; **55:** graphic-bee, colematt; **56:** colematt, yelet, FloWBo; **58:** yelet, FloWBo, Iuliia Kanivets, jehsomwang; **59:** S953554, jehsomwang; **60:** nyamol; **62:** KanKhem; **63:** Armation74.

Sticker Pages: Maria Voronchikhin, Olga_Mallari, colematt, luplupme, MaksimYremenko, JuliarStudio, Irina_Qiwi, ArnyGFX, art-sonik.

Preface

Collins Little Learners, comprising 20 books in all, provides graded, age-appropriate learning opportunities targeting holistic skill development for children in the age group of 3 years to 6 years. Spread across three levels, the series spans **English**, **Maths**, **General Awareness**, **Art and Craft** and **Hindi**. The series comes bundled with a **workbook** for English and a book of **rhymes and stories** for each level. The books use a combination of hands-on activities, frequent recapitulation and child-friendly artwork to facilitate learning.

The early years are the most significant years for growth, development and learning. During this period, children need adequate opportunities for stimulating experiences as these favourably affect various learning abilities. Learning experiences, when chosen and planned properly, fuel the innate curiosity of children, and stimulate their thought processes.

Learning and Development During the Early Years

The goal of learning during the early years is to enhance the all-round development of personality and to inculcate healthy attitudes, life skills and good values. Creative play, well thought-out games and developmentally appropriate activities help children develop the necessary skills to achieve these goals. The skills that children need to acquire can broadly be categorized as:

Motor: Motor skills pertain not only to gross motor activities like walking and jumping but also include fine motor activities like the ability to hold a pencil and write. The development of motor skills is facilitated through several graded activities.

Personal, Social and Emotional: Personal, social and emotional development, necessary for the overall personality development of children, is achieved using a combination of hands-on activities, role plays, and thought-provoking stories and rhymes.

Linguistic: The ability to listen, speak, read and write is critical to success in formal learning environments.

Mathematical: The development of mathematical ideas and concepts is important at this stage since children will continue to build on this learning for most of their years of schooling.

Knowledge and Understanding of the World: This is achieved through activities that enable children to explore their surroundings at home and in the school, and interact with the people that they see around them.

Health and Safety: The idea of health and safety is introduced through activities, stories and retelling of real-life experiences to encourage children to focus on health and well-being.

Creative: Creative skills are developed through a host of activities involving art, craft, painting, roleplays, etc.

Logical and Rational Thinking: Activities that involve simple puzzles, pattern-completion and visual discrimination help children develop logical and rational thinking skills.

The Collins Little Learners Teacher Support Materials

The series includes extensive teacher support in the form of:

Teacher's Resource Books: Comprises stories, activity banks, classroom management and assessment guidelines

Multimedia eBook: Consists of animations, audio, video, slide shows, and interactive exercises

Print-ready Worksheets: Additional worksheets for practice and reinforcement, present in the digital support

Digital Flashcards: Flashcards with embedded sounds and animations for letter formation and letter-sound association

Animations and Songs: Animated stories and rhymes set to tune

Interactive Games: Interactivities to reinforce concepts in a fun and engaging manner

A note on the activity banks (included in the Teacher's Manuals)

The activity banks comprise hands-on activities that will help children develop the requisite skills to attain age-appropriate learning goals. In keeping with Howard Gardener's theory of multiple intelligences, the activity banks include options and variations that enable teachers to provide a planned cycle of experiences and opportunities, which foster each type of intelligence, to every child in their classrooms. The activity banks equip teachers to use the strengths of children as a means of complementing their less developed areas by valuing and nurturing the inherent abilities of the children.

Contents

Walkthrough

Number formation with pattern tracing

Exercise to develop number-value association

Exercise to reinforce number recognition

Pattern tracing for fine motor development

Two-page spreads have been used to introduce numbers. The emphasis is on improving number recognition and developing number–value association through a variety of exercises. The teacher's manual includes stories, which can be used to accomplish these goals.

Pre-numeracy concepts introduced through short rhymes to make it fun for little learners, followed by practice pages for reinforcement.

Activities such as spot the difference help develop visual and observation skills.

The development of logical thinking skills is supported through such pages.

Frequent revision pages have been included to reinforce learning.

Small Rob and Big Ellie

Use cotton buds to colour big Ellie. Paste a sticker on small Rob.

Sing.

Small Rob and big Ellie,
Come to play in the sun.
Small Rob and big Ellie,
Have a lot of fun.

Match the big picture with the small picture.

Tick (✓) the big pictures. Cross (✗) the small pictures.

Milli Mouse Goes Up and Down

Trace the path of Milli mouse. Use [red crayon] to go up. Use [blue crayon] to go down.

Sing.

Hickory dickory dock,
Milli mouse ran up the clock,
The sun went down,
Milli mouse ran down,
Hickory dickory dock!

Match the following.

Colour the pictures.

Billy Goat

Circle the things that are near. Cross (✗) the things that are far.

Sing.

So many pretty flowers I see around me,
Pretty flowers near me,
Flowers I can touch and see.
I see red apples high up on the apple tree,
Red apples so far away from me,
Apples I can only see.

Match the correct pair.

far

near

far

far

near

Colour the thing that is far from the bee.

I Can Colour!

Colour the big picture with . Colour the small picture with .

Colour the correct arrow for each image.

Fun Time

Spot all the differences between the pictures.

Gabby Giraffe and Zilly Zebra

Circle the 🥭 that tall Gabby eats. Cross (✗) the 🥭 that short Zilly eats.

Listen to the story of Gabby and Zilly.

Tick (✓) the tall image.

Lily and Her Ribbons

Sing.

Lily has a ribbon long and red.
Lily never wears her ribbon to bed.
Lily ties it in a bow on top of her head.
Lily's ribbon long and red.

Lily has a ribbon short and blue.
Lily's ribbon is brand new.
Lily's ribbon keeps her hair tidy too.
Lily's ribbon short and blue.

Cross (✗) the short image.

Colour the long snake.

Draw a short ribbon. Colour it blue.

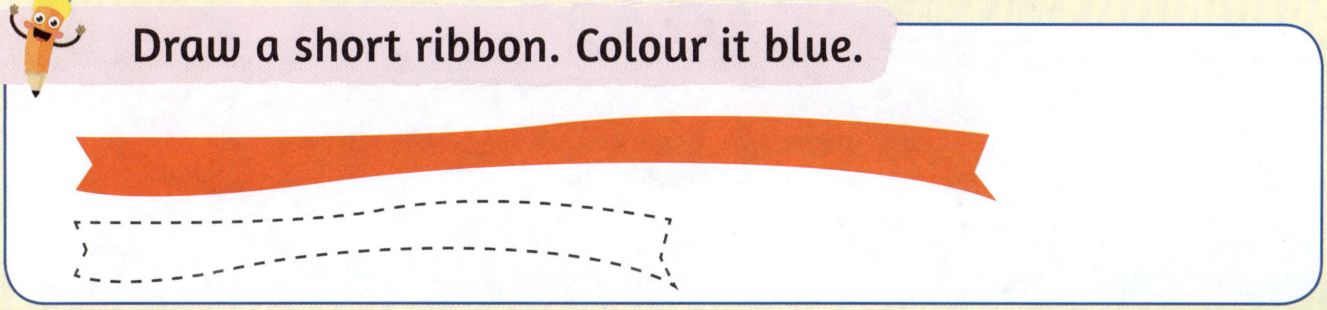

Willy Nilly

Sing.

Willy 'n' Nilly are two fat puppies.
Willy loves cheese, Nilly loves candies.
Willy 'n' Nilly do not like to run.
Willy 'n' Nilly like to sleep in the sun.
Willy 'n' Nilly, you must play,
In the park, every single day.

Colour the fat pup blue and the thin pup red.

Tick (✓) the fat image. Cross (✗) the thin image.

I Know!

Colour the pictures with the colour that matches their height.

Trace the lines to match the correct pictures.

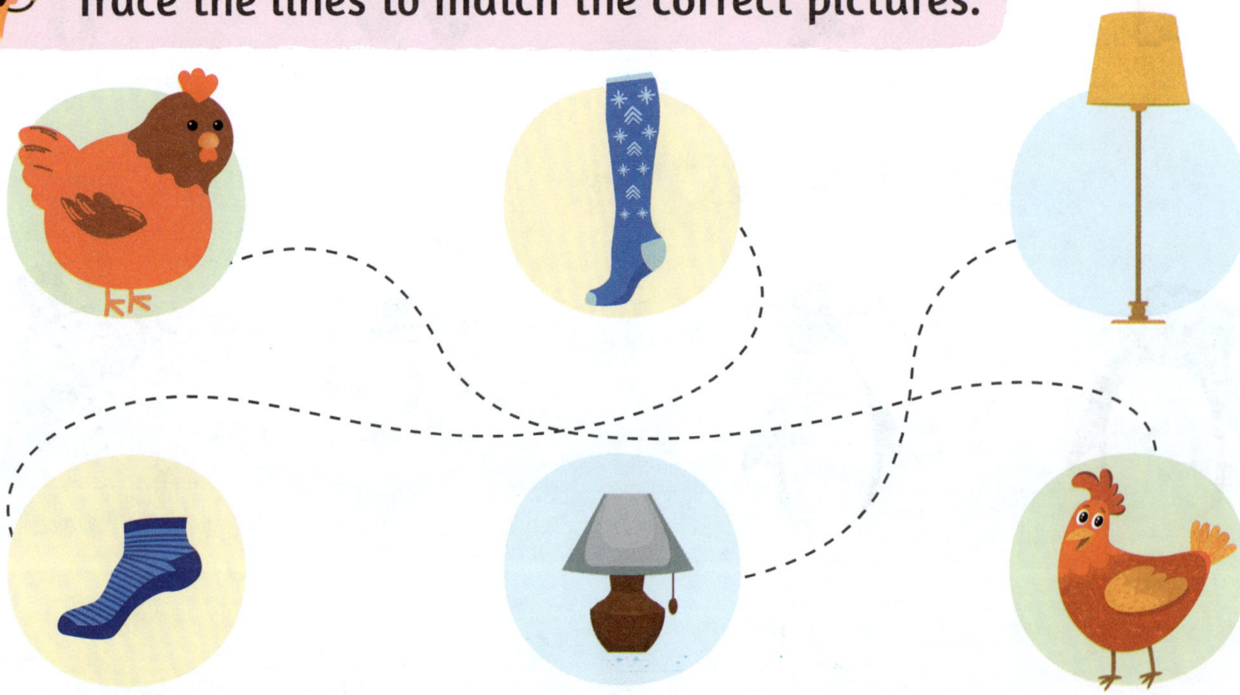

Trace and colour the long picture in yellow.

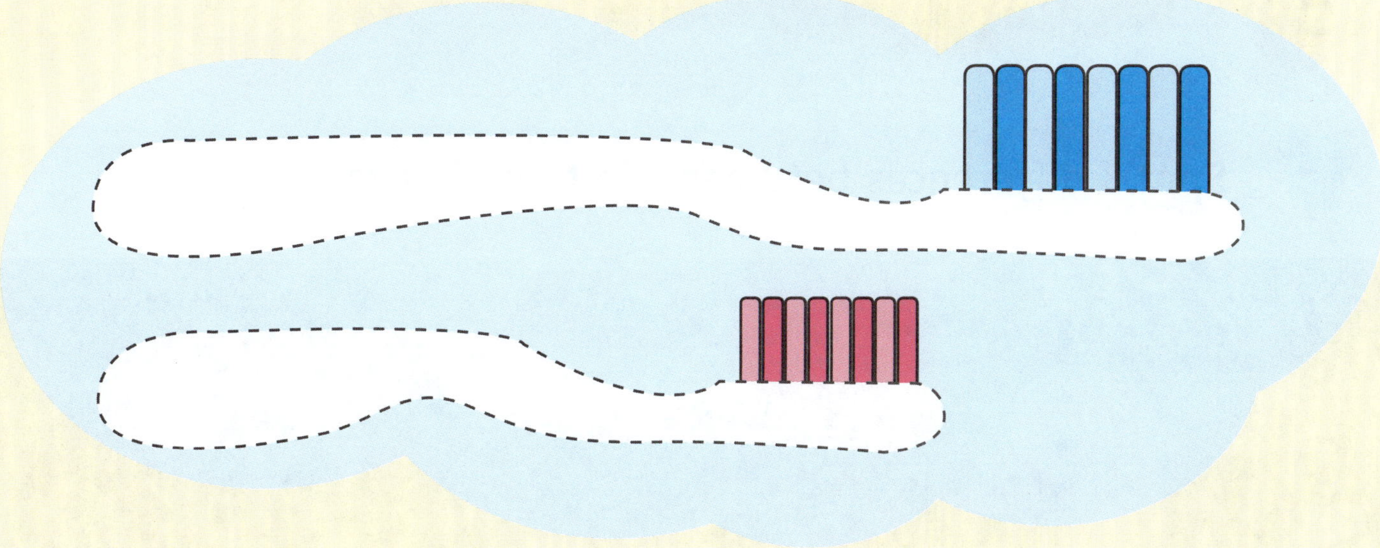

Trace and colour the fat hen in red.

Fun Time

Spot 3 differences between the two pictures.

Cross (✗) the odd one out.

Pattern Tracing

Trace the patterns with your finger.

Trace and complete the picture. Colour the picture.

Shapes

Trace the shape. Paste the matching sticker.

Stick here

square

Tick (✓) the shape that is a square.

Shapes

Trace the shape. Paste the matching sticker.

Stick here

circle

Tick (✓) the shape that is a circle.

Shapes

Trace the shape. Paste the matching sticker.

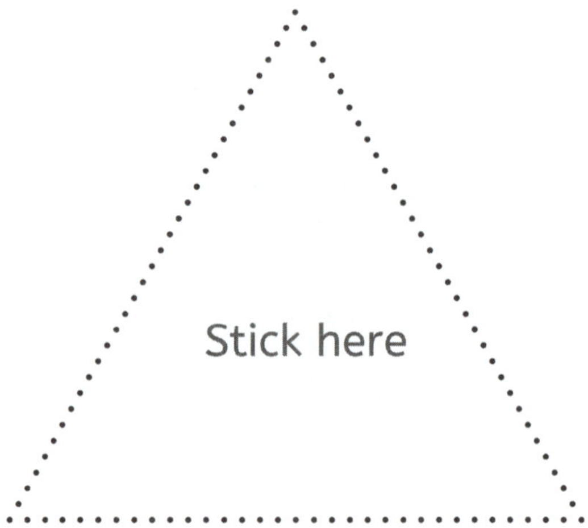

triangle

Tick (✓) the shape that is a triangle.

Paste matching stickers to complete the picture.

Meet Number 1

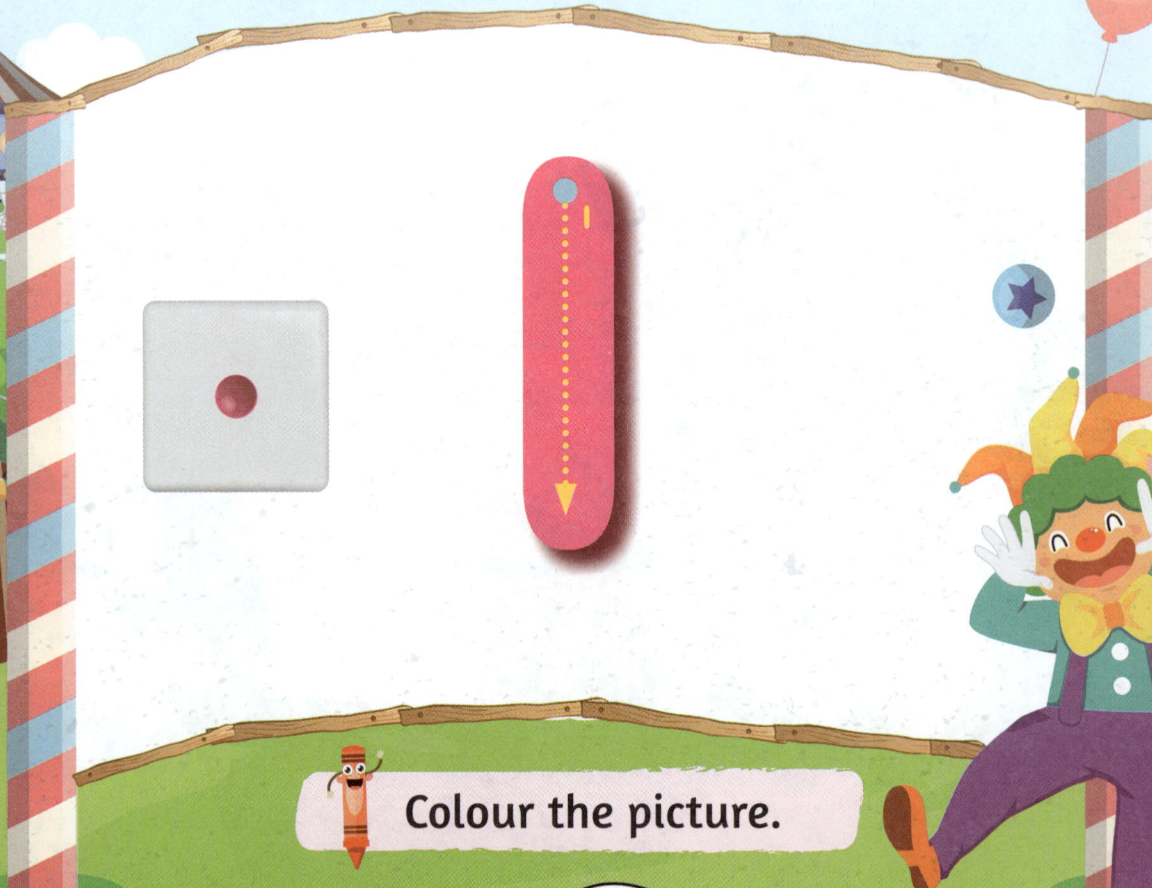

Colour the picture.

Colour the birds with number 1.

Trace with your finger.

SING ALONG

One grey eagle was happy it won

The friendship of number 1.

Meet Number 2

Sponge-print to make patterns on the cows.

Colour **2** animals that are the same.

Trace with your finger.

SING ALONG

Two tan cows began to moo

When they saw number **2**.

Meet Number 3

Tear and paste paper to make the tails of the squirrels.

Colour **3** nuts brown, **2** nuts black and 1 nut green.

Trace with your finger.

SING ALONG

Three thirsty squirrels danced with glee

When they saw number **3**.

Meet Number 4

Colour the hens.

Trace the worms and colour them.

Trace with your finger.

SING ALONG

Four fat hens ran to the door

To welcome the number 4.

Meet Number 5

Thumbprint to complete the bees.

Circle the bees with the number **5.**

Trace with your finger.

SING ALONG

Five busy bees built a hive

When they saw the number **5.**

Tick (✓) the correct picture to complete the pattern.

I Know!

Circle the correct number.

1 3 5

4 2 1

2 3 1

3 2 5

4 5 1

Colour the correct set of pictures.

4

2

5

3

1

Count and match.

1

4

2

5

3

Print your left hand on the page. Count the number of fingers you have in your left hand. Circle the correct number.

Full-Empty

Touch the pictures. Talk about each picture.

Paste stickers of fruits to fill the empty basket.
Colour the basket.

Meet Number 6

Colour the frogs green.

Paste **6** stickers on **6** frogs.

Trace with your finger.

SING ALONG

Six spotty frogs were in a fix

When they landed on the number **6**.

Meet Number 7

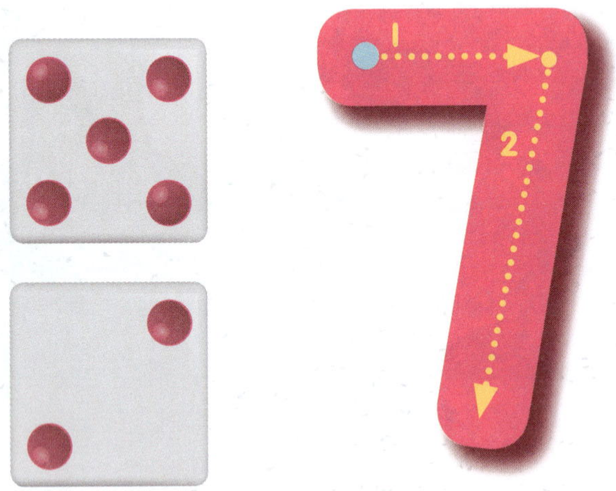

Print with cotton buds to complete the ants.

Trace and colour **7** cubes of sugar for the ants.

Trace with your finger.

SING ALONG

Seven skinny ants playing in the garden
Came across the number **7**.

Meet Number 8

Paste two circles to make **8**.

Cross (✗) **8** birds that look the same.

Trace with your finger.

SING ALONG

Eight elegant swans flew to the gate

To let in the number **8**.

Meet Number 9

Thumbprint to complete the picture.

Paste 9 stickers of water drops for the elephant.

Trace with your finger.

9

SING ALONG

Nine sick elephants felt just fine

As soon as they saw number 9.

Meet Number 0

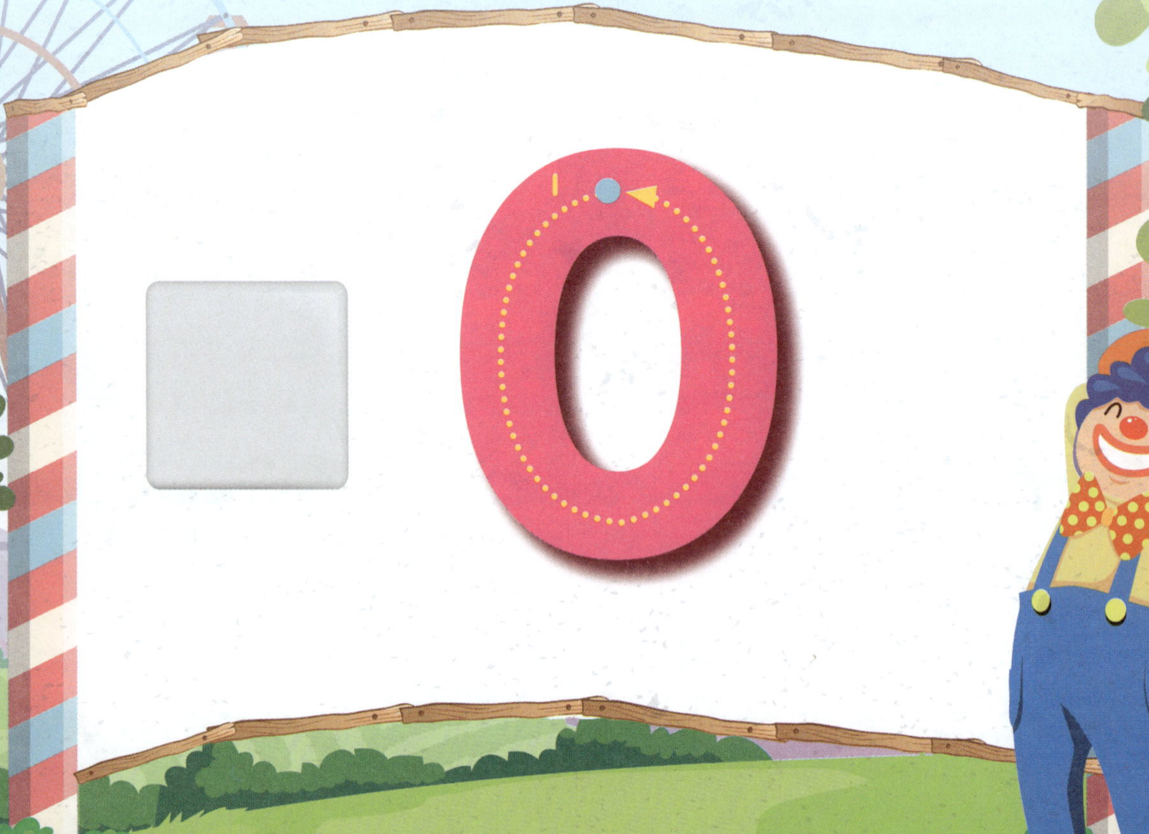

Trace with your finger.

Paste stickers of flowers so that the vase is not empty.
Paint with cotton buds to decorate the vase.

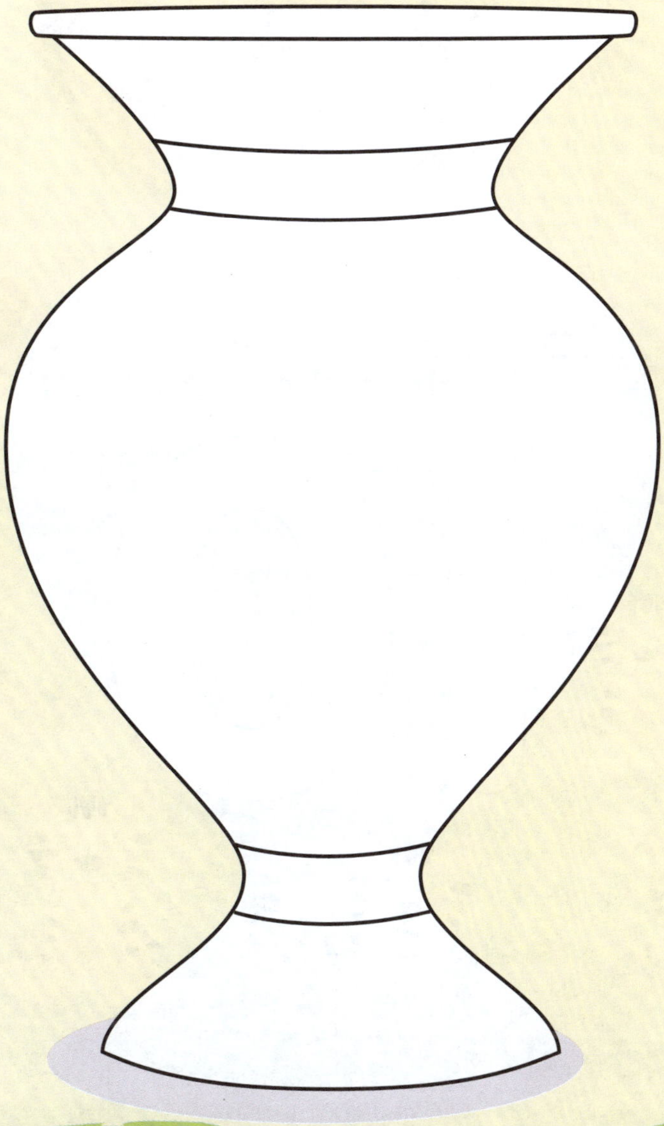

Meet Number 10

Colour the number 10.

Lead 10 tigers to the den.

Trace with your finger.

10

SING ALONG

Ten tawny tigers ran back to the den

They were afraid of number **10**!

I Know!

Match the pictures to the correct number.

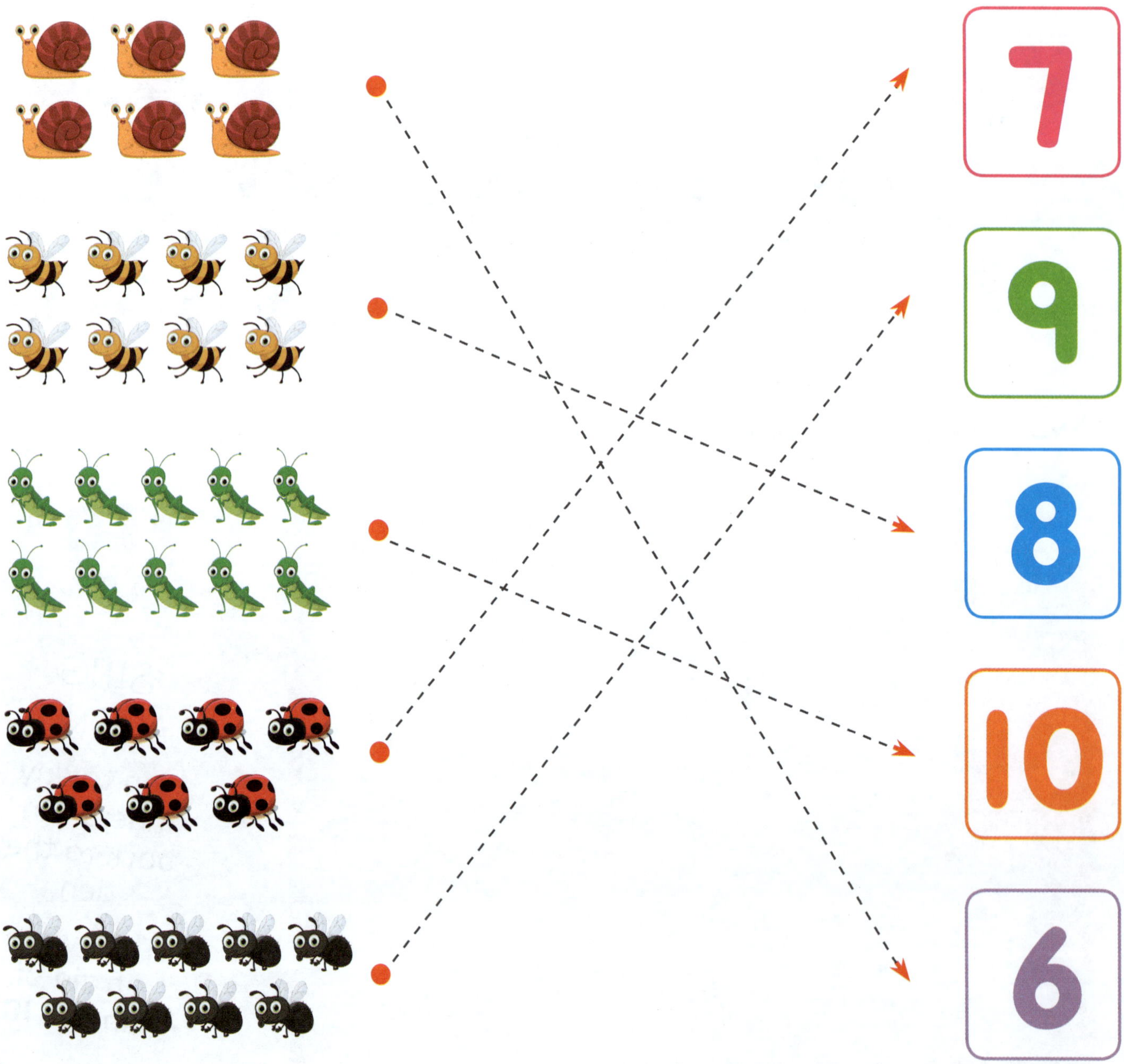

Use a cotton bud to paint the correct number of seeds in the watermelon.

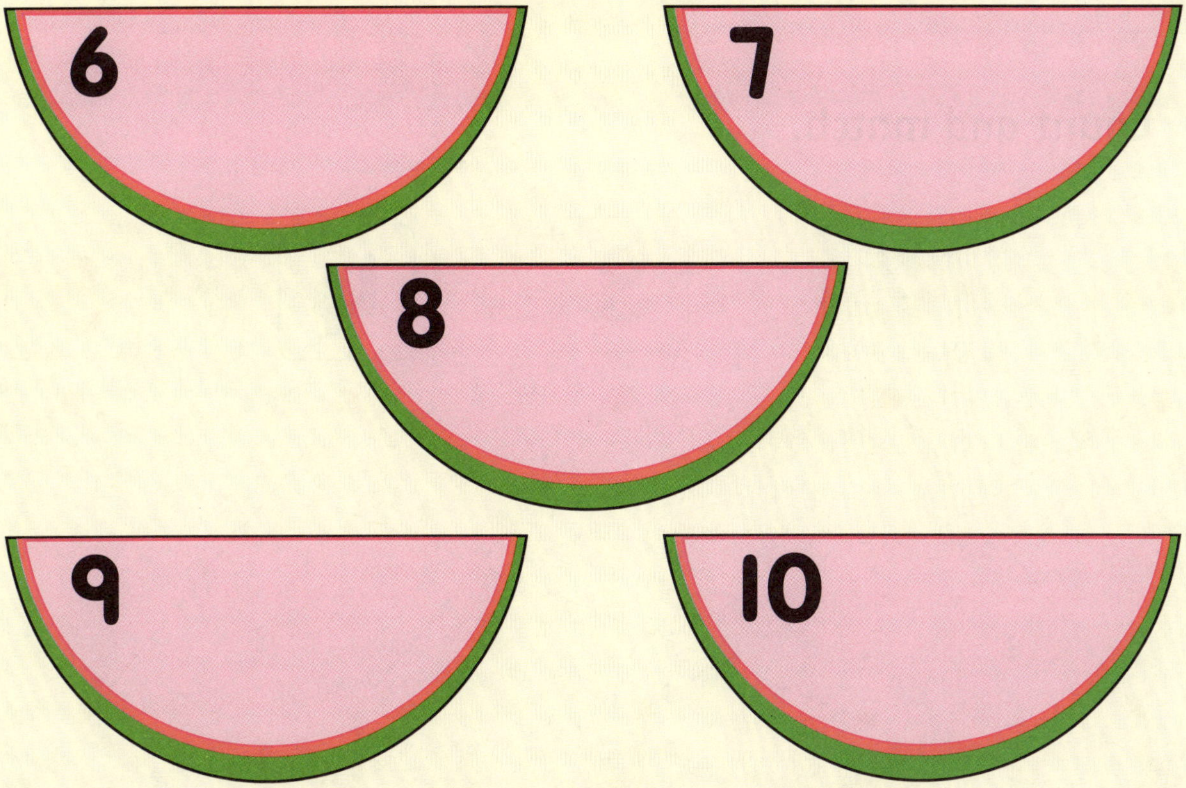

Paste stickers in the correct order.

6 7 8

9 10

I Can Count!

Count and match.

1 2 3 4 5

6 7 8 9 10

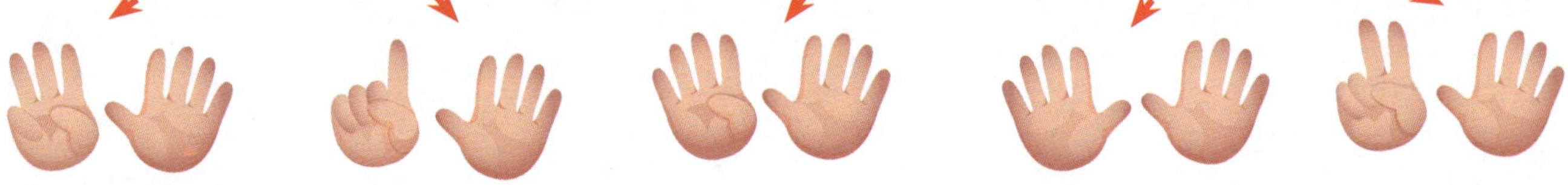

I Can Count!

Paste stickers of the missing numbers.

Tracing Numbers 1-10

Trace the numbers.

Sticker page

Page 8:

Page 28:

Page 29:

Page 30:

Page 31:

Page 47:

Page 49:

Page 52:

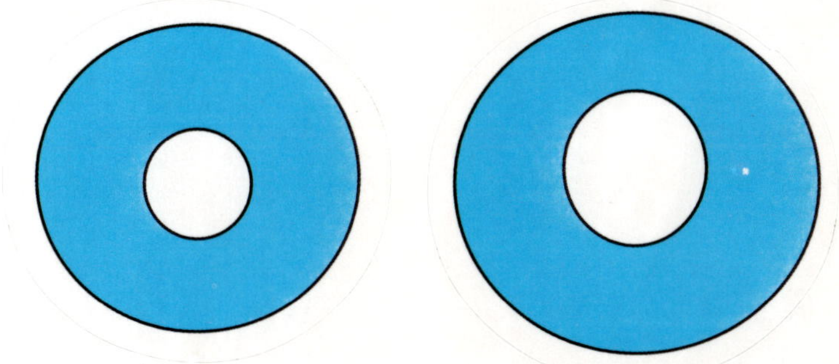

Page 55:

Page 57:

Page 61:

Page 63:

3 4 5 6

7 9 10